From Strategy To Subterfuge

Unraveling The Enigma Of Deception And Concealment In the Realm Of Asymmetrical Conflict and Its Tactics Today

Avery Hambleden

Table of Contents

1 Introduction: The Nature of Deception Chapter 1: Foundations of Asymmetric Warfare

Asymmetric warfare fundamentally differs from traditional conflict. It's characterized by a stark imbalance of power, where a weaker adversary employs unconventional tactics to offset the stronger opponent's advantages. Think of David and Goliath—a classic illustration of this uneven playing field. The weaker party leverages ingenuity, surprise, and often, deception, to compensate for their lack of resources and conventional military might. This inherent imbalance shapes the entire dynamic, creating opportunities and vulnerabilities unique to this type of conflict. Understanding these core differences is paramount to developing effective strategies and countermeasures.

Historically, asymmetric warfare has been a constant throughout military history. From ancient guerilla tactics employed against larger empires to the modern-day use of terrorism, the fundamental principles remain remarkably consistent. Consider the countless examples of smaller forces successfully challenging powerful empires through unconventional methods, often employing cunning deception to achieve strategic objectives far beyond their apparent capabilities. The Roman legions faced this reality repeatedly; so did the British Empire during various colonial uprisings. These historical precedents underline the enduring relevance and adaptability of asymmetric warfare strategies across vastly different technological eras and geopolitical landscapes. Studying these past conflicts provides invaluable insights into the patterns and potential outcomes of current and future conflicts.

The core of asymmetric warfare often lies in deception. It's not merely a tactic; it's frequently a strategic imperative. A weaker force needs to create confusion, misdirection, and doubt in the minds of the stronger opponent. This involves more than just battlefield deception; it extends to the manipulation of information, the creation of false narratives, and the exploitation of vulnerabilities in the enemy's intelligence gathering and decision-making processes. Successful asymmetric campaigns are usually built upon a foundation of skillfully executed deception that can destabilize even the most technologically superior opponent.

Effective deception requires a deep understanding of human psychology. Fear, uncertainty, and doubt are powerful tools, expertly leveraged by asymmetric actors. By skillfully creating these feelings within the opponent's ranks, they can achieve disproportionate results. This understanding of the human element is crucial for designing and deploying deception strategies. The psychological effect of uncertainty and the propagation of false narratives can severely cripple an opponent's ability to act decisively and coherently, turning a seemingly superior force into a hesitant and reactive one. Understanding these psychological dynamics is paramount.

Several methods contribute to successful deception. These

range from classic military ruses—like feigned retreats or strategically placed decoys—to sophisticated information operations. Disinformation campaigns—the deliberate spread of false information—can be incredibly effective in shaping public opinion and influencing political decisions. The modern information age presents even greater opportunities and challenges in this area. It allows the spread of false narratives and propaganda with unprecedented speed and reach. The ability to control and manipulate information flow becomes a key strategic advantage in asymmetric conflict.

This control is deeply intertwined with information operations. These are not merely passive dissemination of misleading information; they're active efforts to manage the information environment, influencing perceptions and shaping the narrative surrounding the conflict. This includes controlling the flow of information to the enemy, creating ambiguity and confusion, and disseminating carefully crafted misinformation to sow discord and uncertainty. This dynamic control of information often proves more effective than brute force, particularly when combating a stronger, more technologically advanced opponent. The ability to skillfully manage the battlefield's informational landscape, therefore, is often the key to asymmetric victory.

The nature of deception in asymmetric warfare is continuously evolving. Technological advancements bring new tools and capabilities, demanding constant adaptation and innovation. The advent of cyber warfare, for instance, has added a whole new dimension to deception. The ability to hack systems, disrupt communication networks, and spread disinformation online creates unprecedented opportunities for manipulation and control, making the digital realm another critical battleground in this ongoing conflict. Adapting to these technological shifts is essential for both sides in the conflict. The ability to stay ahead of the curve—to anticipate and counter the ever-evolving methods of deception—is a critical factor in success.

1.1 Defining Asymmetric Conflict

Asymmetric conflict isn't simply a clash of armies; it's a fundamentally different type of warfare. It's defined by the stark imbalance of power between the combatants, a chasm often bridged by ingenuity and deception rather than brute force. One side, typically a weaker state or non-state actor, leverages unconventional tactics and strategies to offset the superior resources and technological prowess of their opponent. This inherent imbalance shapes the entire conflict, dictating the tactics employed, the strategies pursued, and even the very definition of victory. Think of David and Goliath, but instead of a sling, consider cyberattacks, guerrilla warfare, or the manipulation of information. The disparity isn't just in weaponry; it encompasses everything from manpower and funding to technological capabilities and political influence.

This fundamental imbalance fundamentally alters the rules of engagement. Traditional military doctrines, designed for symmetrical conflicts, often prove inadequate. Successful asymmetric campaigns rely on exploiting vulnerabilities, not through head-on confrontation, but through indirect approaches. Consider a smaller, less technologically advanced force employing sophisticated intelligence gathering, coupled with targeted strikes against critical infrastructure, to neutralize a technologically superior opponent. This subtle but powerful approach exemplifies the core nature of asymmetric conflict.

Understanding this core imbalance is critical. It's not merely a matter of recognizing the differences in resources; it's about grasping the strategic implications of those differences. A stronger state may possess advanced weaponry, but a weaker one might utilize unconventional tactics that exploit the former's vulnerabilities. For example, a terrorist organization might leverage social media to spread propaganda and incite unrest, a method not easily countered by conventional military means. This highlights how asymmetric conflict demands a completely different strategic mindset.

The very nature of asymmetric conflict makes it difficult to define precisely. There is no single defining characteristic; instead, it's a constellation of factors, including the disparity in military capabilities, the use of unconventional warfare tactics, and the exploitation of vulnerabilities within the adversary's political or social systems. A key element is the blurring of lines between war and peace, often characterized by the use of proxies, covert operations, and unconventional methods. This fluidity challenges conventional notions of battlefield and enemy, demanding a broader, more nuanced understanding of the conflict dynamics.

Analyzing historical examples offers invaluable insights. The Vietnam War, for example, demonstrated the effectiveness of guerrilla warfare against a technologically superior force. Similarly, the resistance movements against occupying forces throughout history illustrate how unconventional tactics, such as sabotage and guerilla warfare, can effectively challenge even the most powerful armies. These historical precedents illuminate the enduring nature of asymmetric conflict and the adaptability of unconventional warfare tactics. The key takeaway isn't simply the success of the weaker party; it's the demonstration that a mismatch in resources doesn't automatically equate to inevitable defeat.

Successfully navigating the complexities of asymmetric conflict necessitates a profound shift in strategic thinking. Traditional doctrines focusing on decisive battles and overwhelming force are often ineffective. Instead, strategies must account for the inherent ambiguities and the need for adaptability. This requires a deep understanding of the adversary's capabilities, motivations, and vulnerabilities. Intelligence gathering, therefore, takes on heightened importance, informing the development and execution of tailored strategies designed to offset the adversary's advantages.

Effective strategies often blend military, political, and informational approaches. The objective might not be a conventional military victory, but rather to achieve political objectives through a combination of direct and indirect action. This requires a multi-faceted approach, engaging in actions that may not seem strictly military but nonetheless contribute to the overarching strategic goal. A campaign of economic sanctions, for instance, can cripple an adversary's ability to wage war, just as effectively as a military offensive.

Moreover, it is essential to understand the psychological dimension. Asymmetric warfare often relies on shaping perceptions, both domestically and internationally, to influence the course of the conflict. Propaganda, disinformation campaigns, and psychological operations are crucial elements in shaping public opinion and eroding the adversary's will to fight. These efforts can profoundly impact the conflict's trajectory, often proving as potent as military actions themselves.

In conclusion, defining asymmetric conflict requires moving beyond simple definitions of resource disparity. It's about acknowledging the fundamentally different strategic and tactical landscape it creates. Understanding the unique characteristics—the imbalance of power, the unconventional tactics, the exploitation of vulnerabilities—is the first step towards effective engagement and responses. Analyzing historical examples and understanding the psychological dimensions are crucial for developing effective strategies that address the complexities and ambiguities inherent in this unique form of warfare. The ability to adapt, innovate, and understand the multifaceted nature of asymmetric conflict is paramount for success in this ever-evolving landscape.

1.2 Historical Precedents

The Trojan Horse, a seemingly innocuous gift concealing an army, stands as a potent symbol. Its enduring presence in our lexicon underscores the timeless effectiveness of deception in warfare, a tactic far older than gunpowder or even organized armies. From ancient sieges to modern cyberattacks, the fundamental principles remain remarkably consistent: exploit vulnerabilities, manipulate perceptions, and achieve victory through cunning rather than sheer force. This enduring legacy showcases how understanding history provides invaluable context for analyzing contemporary conflicts.

Consider the historical use of camouflage and concealment. The Roman army, renowned for its discipline and logistical prowess, employed sophisticated strategies for both. Their marching formations, designed not just for tactical maneuvers, but also for concealing their true numbers and intentions, served as a precursor to modern military operational security. Similarly, the Viet Cong's masterful use of the jungle, blending seamlessly with their environment to ambush and evade their larger, more technologically advanced adversaries, demonstrates the enduring power of environmental camouflage.

However, the effectiveness of deception hinges on more than just physical techniques. Psychological manipulation is equally crucial. Sun Tzu's Art of War, a timeless text on military strategy, emphasizes the importance of understanding the enemy's mind, anticipating their actions, and manipulating their perception of reality. He advocated for deception as a fundamental component of warfare, stressing the need to create false appearances and exploit weaknesses in judgment. This principle remains powerfully relevant today in the context of information operations and disinformation campaigns. The way in which narratives can shape public opinion, and influence decisions on battlefields illustrates the enduring human susceptibility to deception.

The historical record offers many examples of the devastating impact that a lack of awareness about deception can have on entire armies. The element of surprise, often achieved through carefully crafted deception, has decided the outcome of countless battles throughout history. From Hannibal's unexpected crossing of the Alps, to the D-Day landings, the ability to mask true intentions and strike when the adversary is least prepared has proven decisive. The element of surprise is not merely a matter of timing; it is a consequence of effective deception, a strategy that requires meticulous planning and masterful execution.

Furthermore, the study of historical precedents allows us to appreciate the evolution of asymmetric warfare itself. While the tools and technologies may change – from catapults to cyber weapons – the core dynamic of a weaker adversary leveraging ingenuity to overcome a stronger opponent remains. The tactics of David and Goliath, a seemingly uneven conflict, serves as a foundational archetype for the study of asymmetric warfare, demonstrating how a weaker party can exploit a superior enemy's weakness through unconventional methods.

Examining past conflicts reveals that successful deception isn't about a single, isolated act. It's a multifaceted approach requiring integration of various techniques over time. The prolonged guerilla campaigns of the past illustrate this, where long-term deception played a decisive role in wearing down the will of a larger, more powerful adversary. The endurance and adaptability of such warfare demonstrate the significance of protracted deception strategies that continuously reshape the battlefield perception.

Moreover, recognizing the historical context of these tactics aids in developing countermeasures. By studying how past deceptions were successful – and how they failed – we can gain insights into identifying and mitigating such threats. A deep understanding of historical precedents enables a more nuanced approach to identifying vulnerabilities, countering manipulative tactics, and developing stronger defense mechanisms. Examining the weaknesses exploited by past deception efforts highlights areas that need to be strengthened in modern strategies.

In conclusion, exploring historical precedents offers an invaluable framework for understanding the dynamics of deception and concealment in asymmetric warfare. From ancient battles to modern conflicts, the core principles remain surprisingly relevant. By studying these historical examples, we can gain a deeper understanding of the psychology of deception, the effectiveness of various techniques, and the importance of adapting to the ever-evolving nature of conflict. This historical perspective is crucial for developing effective strategies for defense and offense in the modern age. Understanding the past is essential for navigating the complexities of the present, and preparing for the challenges of the future.

2 Chapter 2: Deception in Warfare

The human mind, a battlefield in itself, is often the primary target in warfare. Deception thrives on exploiting cognitive biases, manipulating perceptions, and ultimately, controlling the narrative. Understanding this psychological landscape is crucial to both employing and defending against deception. We are not merely dealing with physical maneuvers; we are engaging in a complex game of information and psychological warfare. The effectiveness of any deception hinges on the opponent's belief system, their expectations, and their inherent vulnerabilities. A well-crafted deception leverages these weaknesses, planting seeds of doubt and uncertainty to achieve strategic goals. Consider the impact of a carefully orchestrated disinformation campaign; its reach can be far greater than any conventional military operation.

Military deception encompasses a vast array of techniques, each designed to mislead the enemy about one's true intentions, capabilities, or location. These range from simple ruses, such as dummy installations or misleading radio traffic, to highly complex operations involving the coordinated movement of forces and the dissemination of fabricated intelligence. Feints and diversions, for instance, aim to draw enemy forces away from the intended target, creating opportunities for a decisive strike. Camouflage, a more passive form of deception, hides units or equipment, rendering them less visible to observation. The effectiveness of these techniques depends on their plausibility and integration within a broader deception plan. A successful deception isn't merely a singular event but a carefully orchestrated campaign.

Information operations form the backbone of modern military deception. These encompass a wide spectrum of activities aimed at influencing the information environment, shaping perceptions, and manipulating public opinion. Control of the narrative—what information is available, how it's presented, and when it's released—is paramount. The battle for information supremacy involves not only traditional media but also the digital realm, with social media and online platforms becoming crucial battlegrounds. Disinformation campaigns, designed to spread false or misleading information, are a key component of this struggle. Understanding how disinformation spreads, the psychological mechanisms that make it effective, and countermeasures to combat its influence is increasingly critical in asymmetric warfare. Successfully countering disinformation requires a multi-pronged approach, combining proactive information dissemination with robust fact-checking and media literacy initiatives.

The effectiveness of any deception strategy hinges on meticulous planning and execution. Each component must be carefully considered and integrated into a cohesive whole. Furthermore, deception often demands an element of risk and is not guaranteed to succeed. Sometimes, a carefully crafted deception may fail to achieve its intended effect, possibly leading to unexpected consequences. However, the potential gains associated with successful deception often justify the inherent risks involved. A well-executed deception operation can significantly alter the course of a conflict, achieving strategic objectives with minimal losses.

It is vital to note that deception, while a powerful tool, is not without ethical considerations. The use of deception in warfare raises complex moral questions. The intentional manipulation of information, even for a justifiable military objective, has the potential to cause significant harm. It is imperative to draw a clear distinction between deception employed to achieve military objectives and activities that deliberately mislead to inflict unnecessary harm or suffering on civilian populations. A proper ethical framework is crucial to ensure that deception strategies adhere to international humanitarian law and are applied responsibly. The responsible use of deception demands a constant ethical assessment, ensuring the means justify the ends. The goal is not simply to win at any cost, but to achieve victory while upholding moral principles and respecting the laws of war.

2.1 The Psychology of Deception

Understanding the psychology of deception is crucial in asymmetric warfare. It's not simply about tricks and illusions; it delves into the human mind, exploiting vulnerabilities and biases to achieve strategic aims. Successful deception hinges on manipulating the adversary's perception of reality, leading them to make decisions favorable to your objectives, even if those decisions are based on false information. The effectiveness of deception relies heavily on understanding your opponent's cognitive processes, their decision-making styles, and their existing beliefs and assumptions. A thorough understanding of these factors allows for the crafting of believable and compelling deceptions, maximizing their impact. Failing to account for these psychological factors drastically reduces the potential for success.

We are not talking about mere lies, but rather calculated manipulations designed to exploit cognitive shortcuts and heuristics. People rarely analyze information thoroughly; they rely on mental shortcuts to make quick judgments. Deception tactics effectively exploit this tendency. For example, a carefully crafted disinformation campaign can leverage confirmation bias—the tendency to favor information confirming pre-existing beliefs—to plant false narratives in the minds of decision-makers. The more believable the narrative, the more likely it is to be accepted without critical scrutiny. This explains the power of seemingly credible sources and the importance of carefully constructing a narrative that resonates with the adversary's worldview.

Consider the impact of stress and time pressure on decision-making. Under duress, people are more prone to errors in judgment, making them more susceptible to deceptive tactics. A well-timed feint or diversion can overwhelm an opponent, forcing them to react hastily and make poor decisions based on incomplete or misleading information. Conversely, a deceptive campaign that stretches the adversary's resources and attention weakens their ability to critically assess new information, opening them up to further manipulations. The goal isn't just to deceive; it's to create a situation where the adversary's response is predictably flawed.

The psychology of deception goes beyond individual minds; it encompasses group dynamics and organizational behavior. Organizations, like individuals, have biases, ingrained procedures, and cultural norms. Effective deception targets these group-level vulnerabilities. For instance, exploiting an organization's communication bottlenecks or relying on existing bureaucratic processes can create opportunities to inject false information into the decision-making process. Similarly, cultural biases and ingrained assumptions can be manipulated to make a deception more believable. A successful deception often plays on the adversary's expectations and pre-existing beliefs, creating a false sense of security or certainty.

This is where the art of deception blends with the science of intelligence gathering. Before any deception is attempted, thorough intelligence work is required. You must understand the adversary's strengths, weaknesses, command structures, and decision-making processes. This intelligence informs the choice of deception techniques and helps to predict their likely response. Knowing their information sources and their methods of information verification is crucial. The more you know about your adversary, the more effective your deception will be. Conversely, a failure to gather the necessary intelligence dramatically reduces the probability of a successful outcome. Remember, intelligence is the bedrock upon which successful deception is built.

Ultimately, the psychology of deception in asymmetric warfare is a complex interplay of cognitive biases, emotional responses, organizational dynamics, and information manipulation. Mastering this complex interplay requires a deep understanding of human behavior, coupled with skilled intelligence gathering and creative strategic thinking. It is not just a matter of using clever tricks; it's about exploiting inherent vulnerabilities in the human psyche and leveraging them to gain a strategic advantage. Success in asymmetric conflict often hinges on this subtle art of psychological warfare. This understanding is not just a theoretical advantage; it's a practical necessity for those who seek to thrive in the shadowy realm of asymmetrical combat. In short, success depends on outsmarting your opponent not just militarily, but psychologically.

2.2 Types of Military Deception

Military deception, a cornerstone of asymmetric warfare, takes many forms, each demanding a different approach to understanding and countering it. We're not just talking about smoke and mirrors; it's a complex tapestry woven from psychological manipulation, technological innovation, and shrewd strategic planning. Understanding these diverse tactics is crucial for navigating the treacherous landscape of modern conflict.

One of the most basic, yet still effective, forms of deception involves feints and ruses. These are tactical maneuvers designed to mislead the enemy about your true intentions, drawing their attention away from your main objective. Imagine a military creating a series of dummy convoys, complete with fake vehicles and equipment, to divert attention from a genuine troop movement. The success of such a deception hinges on meticulous attention to detail and the ability to maintain the illusion until the crucial moment of the actual operation. A poorly executed feint can reveal more than it conceals, leaving your forces vulnerable.

Disinformation, a more subtle yet potent tool, involves the deliberate dissemination of false or misleading information. This could range from carefully crafted propaganda campaigns designed to erode enemy morale and sow discord within their ranks to leaked intelligence, strategically manipulated to confuse and misdirect their decision-making processes. The effectiveness of disinformation relies on its plausibility and its ability to reach the intended audience before the truth is known. A successful disinformation campaign can sow chaos and undermine an enemy's confidence long before any actual combat occurs. Consider, for instance, planting false information about troop movements or weapons capabilities to trigger a premature response from the enemy, creating a window of opportunity for a strategic offensive.

Deception also extends to the realm of camouflage and concealment, techniques aimed at physically masking troops, equipment, or installations. While seemingly straightforward, modern camouflage is sophisticated, leveraging advanced materials and technology to blend into the surrounding environment, rendering detection almost impossible. Consider the development of materials that adapt to their surroundings, changing color to match the environment and minimizing their visual signature. Or think about the use of digital camouflage to disrupt detection systems. However, even the most advanced camouflage technologies can be compromised through superior intelligence-gathering, highlighting the interplay between deception and intelligence gathering, a key element of modern warfare.

Another crucial aspect is operational security (OPSEC). OPSEC isn't about hiding your intentions entirely but rather minimizing the information available to your enemy that can be used against you. It's about controlling what is visible to adversaries, from communications to daily routines. Simple measures, such as secure communication protocols or carefully planned movements, can significantly reduce the enemy's ability to gain valuable intelligence. Maintaining rigorous OPSEC is an ongoing process, demanding continuous vigilance and adaptation to counter evolving intelligence-gathering techniques. A single lapse in OPSEC can undo the effectiveness of elaborate deception plans, providing crucial intel for an adversary.

Finally, electronic warfare and cyber deception play an increasingly critical role in modern asymmetric conflict. These techniques employ sophisticated technology to manipulate electronic signals, disrupt communications, and even infiltrate enemy systems to spread disinformation or disable critical infrastructure. The ability to mask your online presence through digital camouflage or intercept and manipulate enemy communications creates a new dimension of deception. The blurring of physical and digital realms necessitates a multi-layered approach to deception, requiring expertise across various domains. Moreover, ethical considerations are paramount, as these sophisticated methods can cause significant collateral damage or even lead to unintended consequences that far outweigh any tactical advantage. These are not just tools of war; they are tools with significant ethical implications. Therefore, understanding their ethical ramifications and consequences is crucial for responsible deployment.

2.3 Information Operations

Information operations are the backbone of deception in modern asymmetrical warfare. They're not merely about spreading misinformation; they're about carefully crafting a narrative that shapes the enemy's perception of reality, influencing their decisions, and ultimately, achieving a strategic advantage. This requires a multi-faceted approach, going far beyond simple propaganda.

Effective information operations hinge on understanding your adversary. What are their information sources? What kind of information are they most likely to believe or distrust? Knowing this allows you to tailor your message for maximum impact. For instance, disseminating information through channels they trust is crucial. Conversely, discrediting their preferred sources can sow doubt and confusion. The goal isn't just to spread lies, but to manipulate the flow of information itself.

Consider the strategic use of "leakage." Deliberately releasing carefully chosen pieces of misinformation, or even genuinely accurate but strategically irrelevant information, can overwhelm the enemy's intelligence apparatus. It can force them to expend resources chasing phantoms while neglecting more critical threats, thereby diverting their attention and resources away from your true intentions. This technique, when executed subtly and precisely, can profoundly impact the battlefield.

Moreover, information operations aren't limited to manipulating the enemy. Equally important is the management of information within your own ranks. Maintaining operational security and ensuring the consistent dissemination of accurate, timely intelligence to your forces are equally critical. Internal cohesion and morale are profoundly affected by the information environment; false narratives or conflicting reports can severely undermine an army's fighting capabilities. This emphasizes the importance of rigorous internal communication and validation procedures.

The digital age presents both unprecedented opportunities and immense challenges for information operations. Social media platforms, instant messaging services, and online forums now serve as powerful tools for spreading disinformation, influencing public opinion, and creating narratives. However, the very transparency of these platforms also creates a susceptibility to counter-narratives and fact-checking, demanding constant vigilance and adaptive strategies. Successfully navigating this digital landscape requires a deep understanding of social media algorithms, community dynamics, and the psychological mechanisms that drive information sharing.

Furthermore, the importance of counter-information operations should not be underestimated. Developing the capability to rapidly identify and debunk enemy disinformation, and to strategically counter their narratives, is vital to mitigating their impact. This requires investing in robust fact-checking mechanisms, developing digital forensics capabilities, and employing skilled analysts capable of quickly identifying false information campaigns and countering them effectively. This is crucial for maintaining credibility and preventing the erosion of trust in official communications.

Finally, remember the ethical considerations. While deception is a legitimate tool in warfare, it's imperative to adhere to certain boundaries. The deliberate dissemination of false information that could lead to direct harm to civilians is ethically problematic and potentially illegal. Establishing clear lines and protocols concerning information operations is crucial, not only for legal compliance but also for maintaining the moral integrity of any military operation. This involves a complex balancing act, ensuring effectiveness while adhering to ethical guidelines and international laws, especially in conflicts where the lines between civilian and combatant populations are blurred. The long-term consequences of actions must be carefully weighed against short-term tactical gains. Therefore, a robust ethical framework is essential for the responsible use of information operations.

3 Chapter 3: Concealment Techniques

Camouflage and concealment are fundamental, yet often overlooked, aspects of asymmetric warfare. Effective camouflage isn't just about blending in with the environment; it's a multifaceted approach requiring careful consideration of the visual, auditory, thermal, and even electromagnetic signatures an entity presents. This means selecting appropriate materials, adapting to changing light conditions, and understanding the sensory capabilities of the adversary. Simple techniques, like using natural materials to break up outlines, are crucial, especially for personnel operating in close proximity to the enemy. Remember, even the slightest imperfection in concealment can lead to detection. Advanced techniques, including specialized paints and fabrics that absorb or scatter light, and even the use of active camouflage that dynamically adjusts to the surroundings, require dedicated research and significant investment. Mastering this crucial skill is paramount for any participant in asymmetric warfare.

Operational security, often abbreviated as OPSEC, is the cornerstone of any successful concealment strategy. It encompasses all actions taken to protect information, actions, and intentions from the enemy. This extends beyond simply controlling classified information; it includes carefully managing all aspects of an operation, from planning and rehearsals to communication and movement. A thorough OPSEC plan needs to account for human factors; even a seemingly insignificant detail, like a carelessly discarded piece of trash, can compromise an entire operation. Strict adherence to communication protocols, secure data handling procedures, and regular risk assessments are key components of a robust OPSEC program. Remember that the weakest link can compromise the entire chain. Effective OPSEC is a continuous process that requires constant vigilance and adaptation.

Network security in the context of asymmetric conflict is vital given the increasing reliance on communication and data networks. This includes protecting computer systems, communication networks, and data from unauthorized access, use, disclosure, disruption, modification, or destruction. This requires robust cybersecurity measures such as firewalls, intrusion detection systems, and encryption, but it is not just technological solutions that matter. Human factors play a critical role in network security. Strong password policies, employee training on cybersecurity best practices, and a culture of security awareness throughout an organization are paramount in prevention. Regular security audits and vulnerability assessments are crucial to maintaining a secure network. Furthermore, developing contingency plans to respond to and recover from cyberattacks is essential in today's interconnected world.

Physical security focuses on protecting personnel, facilities, and equipment from unauthorized access and damage. This goes beyond simple perimeter security and includes measures like access control systems, surveillance technologies, and physical barriers. However, effective physical security is more than just technology; it incorporates risk assessment, threat analy-

sis, and well-trained security personnel. The design of buildings and infrastructure can incorporate security features, reducing vulnerabilities. Furthermore, regular security drills and training can equip personnel to respond effectively to security breaches. Remember, a multi-layered approach, combining both technological and human security measures, is the most effective way to secure assets in a hostile environment.
Stealth technologies represent the cutting edge of concealment, drawing on advanced scientific and engineering principles to minimize an entity's detectability. These include techniques to reduce radar cross-section, infrared signature, acoustic emissions, and even visual signatures. The development and implementation of these advanced technologies often require substantial resources and technical expertise. Furthermore, the constant arms race between those seeking to conceal themselves and those trying to detect them creates an ever-evolving landscape of techniques and counter-techniques. However, even the most sophisticated stealth technologies are not foolproof; they are only as effective as the operational security measures that support them. The interplay between technology and human factors is still essential.

3.1 Camouflage and Concealment

Effective camouflage and concealment are paramount in asymmetrical warfare. Success often hinges on the ability to surprise and overwhelm the opponent, a feat achievable only through meticulous planning and execution of concealment strategies. These aren't simply about hiding; they're about manipulating the enemy's perception of reality.
Understanding the environment is critical. A successful camouflage scheme blends seamlessly with its surroundings. This

requires detailed knowledge of terrain, vegetation, and even weather patterns. Consider, for example, a sniper team operating in a desert environment. Their clothing, equipment, and even their position must blend with the sand, rocks, and sparse vegetation. Failure to achieve this perfect blend can mean exposure and defeat. Similarly, in an urban setting, the camouflage would need to adapt, using building materials, shadows, and even civilian attire as a means of concealment. The specific methods change dramatically with the environment.

Beyond blending in, consider misdirection. Creating false impressions through carefully placed decoys or misleading signals can draw enemy attention away from actual troop movements or operational areas. This may involve the strategic placement of dummy vehicles, equipment, or even the dissemination of false information through various channels. This tactic, historically effective, remains crucial today. Imagine a small, lightly armed group attempting to infiltrate an enemy base. By using decoys and diversions, they can create the illusion of a larger, more threatening force, distracting the enemy and buying them valuable time and strategic advantage.

Technological advancements have revolutionized camouflage and concealment. Advances in materials science have led to the development of advanced camouflage fabrics that can adapt to different environments and alter their appearance to match their surroundings. These are not just simple patterns; they're dynamic systems that change and react to light, temperature, and even movement. Furthermore, the use of thermal camouflage significantly reduces the enemy's ability to detect presence using infrared technology. Such technology offers an unprecedented level of protection.

But technology is only as effective as its operators. Human error can negate even the most advanced systems. Therefore, rigorous training in camouflage and concealment techniques is essential. This training must cover not only the technical aspects of camouflage but also the psychological aspects of deception. It needs to incorporate practical exercises in diverse terrains and conditions and must be adapted to the individual needs and weaknesses of each soldier. It is not enough to possess advanced materials or equipment; those materials must be used correctly.

Operational security (OPSEC) plays a vital role in maintaining concealment. This involves identifying and mitigating potential vulnerabilities that could expose your operations. Thorough planning, secure communications, and careful control of information are critical to preventing leaks that could compromise your plans or compromise the security of your forces. Maintaining OPSEC isn't a one-time event but an ongoing process that needs to be carefully managed and adapted to changing circumstances. A single lapse in OPSEC can have devastating consequences.

Physical security complements OPSEC, focusing on the protection of personnel, equipment, and facilities. This involves implementing physical barriers, access controls, and surveillance systems. However, physical security is not just about technology; it also involves training personnel to recognize and respond to potential threats. This includes awareness of their surroundings, being vigilant for suspicious activity and responding swiftly and effectively to any intrusions. Effective physical security requires a combination of technology and human vigilance.

Network security is crucial in the modern digital battlefield. Protecting sensitive information from cyber threats is as vital as protecting physical assets. This involves employing firewalls, intrusion detection systems, and strong passwords to safeguard data and communications. It also includes educating personnel about online security best practices and potential threats. It's a continuous battle against sophisticated attacks.

In the realm of asymmetrical conflict, camouflage and concealment aren't just tactical tools; they're strategic necessities. Mastering these techniques – from understanding environmental nuances to deploying the latest technologies, and from diligent OPSEC to robust physical and network security – provides a considerable advantage. Remember, deception isn't about outright lying; it's about controlling information and manipulating the enemy's perception of reality. This demands precision, adaptability, and relentless vigilance.

3.2 Stealth Technologies

Reducing a target's ability to detect, locate, and track assets is paramount in asymmetrical warfare. Stealth technologies, therefore, are not merely supplementary but fundamental to achieving a decisive advantage. They offer the potential to disrupt enemy operations, gather intelligence undetected, and ultimately dictate the terms of engagement. Mastering these techniques can significantly alter the balance of power.

Effective camouflage is more than just blending in with the environment; it's about understanding the enemy's sensor capabilities and exploiting their limitations. This involves meticulous planning and the use of advanced materials designed to absorb or deflect radar, infrared, and visual detection systems. Consider the development of metamaterials—engineered materials exhibiting unusual electromagnetic properties—which can offer superior cloaking abilities compared to traditional camouflage. Their ability to manipulate electromagnetic waves allows for the creation of materials that effectively render objects invisible to various detection methods.

Low-observable aircraft, for instance, use a combination of techniques, from specially designed airframes that scatter radar waves to coatings that absorb infrared radiation. The F-22 Raptor, a prime example, uses such advanced technologies, enabling it to maintain a degree of surprise and tactical superiority. But low-observable technology is not just limited to aircraft; the application extends to naval vessels, ground vehicles, and even smaller, more portable equipment like drones. Furthermore, understanding the limitations of the adversary's surveillance capabilities, and exploiting these weaknesses through creative concealment and technological means, is essential.

Beyond physical concealment, consider the significance of masking electronic signatures. Sophisticated signal jamming can disrupt enemy communication and surveillance systems, creating confusion and hindering their ability to track movements. However, this demands a deep understanding of the adversary's communication systems and their vulnerabilities. The development of advanced encryption and secure communication protocols becomes critical in maintaining operational security in the face of sophisticated signal intelligence capabilities. Remember, a successful stealth operation requires not only masking physical presence but also securing the flow of vital information.

Furthermore, the integration of stealth technologies with other tactics, such as deception and misinformation campaigns, significantly enhances their effectiveness. By creating a false impression of one's capabilities and intentions, one can lure the adversary into making predictable and, therefore, exploitable mistakes. This synergistic approach requires careful coordination and a deep understanding of psychological warfare, blending technological prowess with the art of deception. The complexity is substantial, and success relies on meticulously coordinating various elements of the plan.

Ultimately, developing and employing stealth technologies effectively requires a significant investment in research and development, advanced training for personnel, and a deep understanding of the adversary's technological capabilities and tactics. This investment pays dividends in asymmetric conflict, where a well-executed stealth operation can provide a disproportionate advantage against a superior force. It's a sophisticated game of technological hide-and-seek that demands constant innovation and adaptation to maintain the upper hand. The ultimate aim is to neutralize the enemy's ability to effectively oppose one's actions, and only through a comprehensive and sophisticated approach to stealth can this be realized.

The future of stealth likely lies in the further development of adaptive camouflage systems, which can dynamically adjust to changing environmental conditions and enemy sensor capabilities. Imagine materials that can seamlessly transition between different appearances, rendering objects virtually undetectable, regardless of their surroundings. This dynamic adaptation would greatly increase the effectiveness of stealth technology, enabling the operator to effortlessly integrate with the environment and evade detection even under close scrutiny. This would require innovative research into advanced materials science, algorithms, and sensor technology.

It's crucial to consider the ethical implications of employing stealth technology. While providing significant tactical advantages, the ability to operate undetected also carries the potential for misuse. Maintaining transparency and adherence to the rules of engagement is crucial. Stealth should not be a tool for covert operations that violate international law or ethical principles. The development and deployment of these capabilities demand thoughtful consideration of their ethical and legal implications to ensure their use aligns with established moral codes and international standards. This underscores the necessity of establishing robust guidelines and regulations to oversee the use and development of these technologies.

The development and deployment of stealth technologies represent a significant shift in the landscape of asymmetrical conflict. These technologies offer the potential to alter the strategic equation, granting a decisive advantage to those who can master their application. However, this advantage comes with significant responsibility, and the ethical and strategic implications of these technologies must always be carefully considered. Ultimately, a thoughtful and balanced approach to the development and deployment of stealth technologies is crucial for ensuring their effective and ethical application within the context of modern warfare. The future battlefield will be defined, in no small part, by the capacity of combatants to effectively utilize these advancements.

3.3 Operational Security

Operational security, or OPSEC, isn't just about secrecy; it's about controlling information. It's the proactive process of identifying, controlling, and protecting sensitive information to prevent adversaries from gaining an advantage. Think of it as a layered defense, a series of carefully constructed barriers designed to keep your intentions, plans, and capabilities hidden. Failing to maintain strong OPSEC can lead to mission failure, compromised assets, and even loss of life. Consider the potential consequences before you act.

Effective OPSEC begins with a thorough understanding of your adversary. Who are they? What are their capabilities and intentions? What information are they most likely to seek? By carefully analyzing these factors, you can tailor your security measures to address the specific threats you face. This is the groundwork. Without it, all other security efforts are undermined. Detailed analysis is non-negotiable. This includes not only their overt actions but also their past behavior and any known intelligence indicating their likely approaches.

Next, you must identify your critical information. This isn't just about classified documents; it encompasses any data that, if compromised, could be used against you. This might include your operational plans, training schedules, communication patterns, personnel details, or even seemingly innocuous information that, when pieced together, could reveal a larger picture. Consider every possible piece of information, regardless of how insignificant it may seem on its own. The power of OPSEC lies in the holistic approach.

Once you've identified your critical information, you need to develop and implement control measures. This might involve limiting access to sensitive data, using secure communication channels, implementing robust physical security measures, and employing deception techniques to mislead adversaries. It is crucial to remember that multiple layers of security are more effective than any single measure. Each layer should strengthen the overall security posture, and every level needs constant evaluation and adjustment as the threat evolves.

Communication security is a critical element of OPSEC. Avoid using easily intercepted communication methods, especially when discussing sensitive information. Always use encrypted channels whenever possible, and be mindful of the metadata associated with your communications, such as timestamps, locations, and participants. Even seemingly innocuous details can reveal critical information about your operations. Think of your communications as a physical space that needs to be guarded against intrusion. Use techniques like using dead drops and avoiding obvious patterns.

Physical security is another vital aspect of OPSEC. Secure your facilities and equipment against unauthorized access, and regularly review your physical security measures to identify and address any vulnerabilities. Implementing strong access control, implementing surveillance systems, and establishing clear security protocols can significantly contribute to reducing threats. Ensure your team knows its role and responsibilities in physical security, and consistently reinforce these measures through drills and trainings. Training is not a one-time event but an ongoing process.

Finally, and perhaps most importantly, maintain a culture of OPSEC within your organization. Educate all personnel on the importance of information security and provide them with the training and tools they need to protect sensitive data. Remember, a strong OPSEC program is not just about procedures and policies; it's about a shared commitment to protecting information and maintaining operational security. The more your team understands the importance of OPSEC, the better they will be at maintaining security. This requires leadership engagement and continual reinforcement. A consistent approach is paramount.

Remember, OPSEC is an ongoing process, not a one-time event. Regularly review and update your OPSEC procedures to address evolving threats and vulnerabilities. Stay informed on the latest techniques and technologies, and adapt your security measures accordingly. This is vital. The landscape of threats is dynamic. What works today might not work tomorrow. Constant vigilance is a necessity, not a luxury.

The best defense is a layered approach that incorporates many different strategies, working together in concert. The individual techniques are only as effective as the training that makes it effective. Every detail matters. By rigorously applying OPSEC principles, you significantly increase your chances of achieving your objectives while mitigating risk. Therefore, strong OPSEC is an essential component of success in any operation. It is not merely about avoiding failure, but achieving consistent success. Without it, success is entirely contingent.

3.4 Network Security

Network security forms a critical layer in the overall concealment strategy within asymmetric warfare. It's not merely about protecting data; it's about safeguarding the very operational heartbeat of a clandestine operation, preventing leaks that could expose your position, plans, or personnel. Consider a simple example: a compromised communications network can betray the location of your forces, turning a strategic advantage into a catastrophic vulnerability. This isn't theoretical; it's a reality in modern conflicts.

Robust network security begins with a thorough understanding of potential threats. These extend beyond the obvious cyberattacks; insider threats, poorly configured systems, and even physical access to equipment can all compromise your network. A comprehensive risk assessment, tailored to your specific operational environment and technological capabilities, is the foundational first step. This assessment should identify vulnerabilities and prioritize mitigation efforts, directing resources where they're most needed. This prevents wasting time and money on unnecessary security measures while ensuring critical systems remain protected.

Implementing a layered security approach is crucial. This means employing multiple layers of protection, working in concert to prevent any single point of failure. Firewalls, intrusion detection systems, and data encryption are all essential components. However, relying solely on technological solutions is insufficient. Human factors play a significant role; rigorous security protocols, comprehensive employee training on safe practices, and fostering a culture of security awareness are vital components to successful network security. Without a well-trained team, even the most sophisticated technology is vulnerable. Regular security audits and penetration testing can help identify vulnerabilities before they can be exploited by an adversary.

Remember that network security isn't a static process. The threat landscape is constantly evolving, with new vulnerabilities and attack vectors emerging regularly. Continuous monitoring, updating of security software, and adaptation of security protocols are essential for staying ahead of the curve. This requires a proactive approach, not a reactive one. Instead of waiting for an attack to occur, regularly scanning for vulnerabilities and implementing patches allows for timely responses and preventative action.

Beyond technical aspects, consider the human element within your network security strategy. Social engineering attacks, where adversaries manipulate individuals to gain access to sensitive information, are increasingly prevalent. Educating personnel about phishing attempts, malicious software, and other social engineering tactics is just as crucial as technical security measures. Regular training exercises can simulate real-world threats, helping your personnel to develop the skills and awareness needed to recognize and respond to potential threats effectively. This proactive training can significantly reduce vulnerabilities associated with human error and a lack of awareness.

Data encryption, a crucial element often overlooked, adds another critical layer of protection. Encrypting sensitive data both in transit and at rest significantly hampers the ability of adversaries to access information even if they do manage to compromise your network. However, strong encryption protocols must be accompanied by careful key management to prevent unauthorized access. Remember that even the most secure system is useless if the keys are compromised. This involves regular key rotation, secure key storage, and robust access control policies. Finally, incident response planning is paramount. Even with the best security measures in place, breaches can still occur. A well-defined incident response plan can help to minimize the damage and speed up recovery time. This plan should outline clear steps for identifying, containing, eradicating, recovering from, and learning from security incidents. The ability to react effectively to an incident can limit the damage, protecting critical information and maintaining operational capability. Regular practice drills and simulations enhance preparedness and team cohesion, ensuring a swift and effective response. The goal isn't merely to prevent breaches, but to minimize their impact when they do occur. A well-structured response plan will determine the success of operations under attack and the effectiveness of network security as a whole.

3.5 Physical Security

Physical security, in the context of asymmetric warfare, goes beyond simple locks and alarms. It's a multifaceted approach that integrates with operational and network security to form a robust defense against infiltration and sabotage. Consider a clandestine operation; effective physical security is crucial for protecting sensitive equipment, personnel, and information. A breach could not only compromise the mission itself but also expose valuable intelligence, potentially altering the strategic landscape considerably.

Think of the seemingly innocuous details. Proper perimeter security, including fencing, lighting, and surveillance systems, creates the first line of defense. This isn't just about deterring casual intruders; it's about slowing down and alerting you to more determined adversaries who may be planning more sophisticated attacks. Sensors, strategically placed, can detect movement, vibration, or even changes in the electromagnetic field. These systems, when properly integrated and monitored, provide early warning of potential breaches, giving you crucial time to react.

Access control is paramount. This encompasses everything from keycard systems and biometric authentication to careful vetting of personnel. Restricting access to sensitive areas based on a 'need-to-know' basis is crucial; overly permissive access dramatically increases vulnerability. It's also vital to establish robust procedures for tracking and managing access credentials, preventing unauthorized duplication or misuse of keys and access cards. Routine audits are a necessary component of such physical security protocols; regularly reviewing and assessing your security arrangements allows you to identify weaknesses and correct them before they're exploited.

Beyond the technical measures, human factors play a pivotal role. Employee training is essential, and it's not just about teaching them how to use security systems. It's about fostering a culture of security awareness. Employees should be trained to identify suspicious activities, report anomalies, and understand the potential implications of security breaches. This includes understanding social engineering tactics, phishing scams, and the importance of secure communication channels. Regular security drills and simulations, mimicking real-world scenarios, help to prepare your team for unexpected events and to ensure they respond efficiently.

The physical security of communication networks is equally critical. Protecting critical infrastructure from physical attack, such as severing cables or disabling satellite uplinks, is essential in a conflict environment. This requires measures such as redundancy, geographically dispersed infrastructure, and hardened facilities. This means investing in robust infrastructure, employing multiple communication pathways, and using encryption to protect data even if it is intercepted. Remember, the physical security of your communication systems is directly linked to the integrity of your operational security; a breach in one compromises the other.

Furthermore, remember the human element when considering physical security. Regularly reviewing your security protocols with your team to discuss vulnerabilities and potential threats is a necessity. This collaborative approach helps identify blind spots and ensures that everyone is aware of their responsibilities in maintaining physical security. Don't underestimate the importance of teamwork; a collective effort strengthens your overall security posture considerably. Regular staff meetings are not only useful to identify potential threats; they build awareness and enhance team cohesion, thus making physical security more effective.

Finally, remember that physical security is a dynamic process. Threats evolve constantly, and what worked yesterday may not be sufficient tomorrow. This necessitates ongoing vigilance, adaptation, and a commitment to continuous improvement. Regular audits and security assessments are not just a formality; they are a vital aspect of ensuring that your physical security measures remain effective and appropriate to the evolving threat landscape. You need to constantly assess potential threats, adapt your approach, and invest in the necessary technologies and training to maintain a strong physical security posture. Don't underestimate the ongoing commitment needed for success.

4 Chapter 4: Cyber Deception

Cyber deception isn't just about hacking; it's a multifaceted strategy demanding careful planning and execution. Think of it as a chess game played in the digital realm, where every move, every byte of data, is carefully orchestrated to mislead your opponent. Success depends on understanding your adversary's digital landscape, their vulnerabilities, and their likely responses. It requires a deep understanding of their networks, their systems, and their procedures.

Creating convincing digital camouflage is crucial. This isn't simply hiding your IP address; it involves crafting a believable digital persona, one that blends seamlessly into the target's environment. Consider a scenario where you need to infiltrate a specific organization. You wouldn't simply create a fake profile; you'd meticulously research the organization's culture, its employees' online habits, and its communication styles. Only then could you create a believable digital identity. This requires extensive research and meticulous attention to detail.

Fake websites, mimicking real government or corporate sites, are a powerful tool. These 'honey pots' can lure attackers, revealing their tactics, and offering valuable intelligence. However, their effectiveness hinges on their realism. A poorly designed website will raise suspicions and be easily identified as a deception. The key lies in attention to detail—even the smallest errors can undermine the entire operation. Think about using genuine-looking graphics, incorporating realistic user feedback, and ensuring that the website functions flawlessly.

One powerful technique involves creating false data trails. This involves strategically placing misinformation within your own network or manipulating data streams to lead attackers down a rabbit hole. The goal isn't to hide entirely; it's to guide the adversary toward a predetermined outcome, diverting their attention from your true objectives. It's a delicate balance—you must create a deception that's believable yet doesn't inadvertently reveal your real operations. It's a high-stakes game of misdirection.

Deploying honeypots is a critical aspect of cyber deception. These are decoy systems, designed to attract attackers and gather information about their methods and capabilities. However, their effectiveness relies heavily on the realism and placement of these systems. Poorly designed or strategically positioned honeypots can be easily identified and bypassed. Consider the layout of the honeypot within the overall network architecture to maximize its effectiveness. Furthermore, continuous monitoring and analysis of the gathered information are crucial for extracting actionable intelligence from the operation. The information gleaned from honeypots should be carefully analyzed and used to refine your defensive strategies. Remember, a good attacker will test and probe your security measures before launching a full-scale attack.

The challenge lies not just in creating the deception, but also in managing the information obtained through it. Analyzing the data collected from these deceptive strategies requires sophisticated techniques and skilled analysts. You're dealing with large volumes of data, often masked and fragmented, requiring significant processing power and expertise to interpret effectively. Moreover, you need to ensure the integrity of the information, verifying its authenticity and distinguishing it from deliberate misinformation planted by the adversary. The ability to effectively interpret the data gathered is essential for extracting valuable intelligence.

Remember, cyber deception isn't a one-size-fits-all solution. The strategies you employ must be tailored to the specific adversary, their capabilities, and their goals. Generic approaches are unlikely to be effective against sophisticated adversaries. Thorough reconnaissance is essential, forming the bedrock of any successful deception operation. Understanding your enemy's strengths and weaknesses is critical to choosing the right techniques. This detailed understanding informs the development of a deception strategy that's highly targeted and effective.

The ethical considerations are paramount. While cyber deception can be a powerful tool, its use requires careful consideration of its potential consequences. Ensure your actions conform to legal and ethical guidelines. Unintended escalation is a possibility, and it is crucial to account for the various consequences your deceptive strategies may have. Remember, while deception can be a valuable tool, irresponsible use can have far-reaching repercussions, damaging trust and undermining international relationships. Always strive to maintain the moral high ground while effectively deploying this strategy. The ultimate goal is to maintain a secure and reliable digital infrastructure, while adhering to a strict ethical compass.

4.1 Cyber Warfare Tactics

Denial-of-service attacks, a cornerstone of cyber warfare, flood networks with traffic, crippling online services. Imagine a coordinated assault targeting a nation's power grid; the ensuing blackout could cause widespread chaos and cripple essential infrastructure. This isn't science fiction; it's a real threat, demonstrating the devastating potential of cyber warfare's simplest tactics.

Sophisticated phishing campaigns, far beyond simple email scams, exploit vulnerabilities in human psychology. These attacks often target specific individuals, crafting personalized messages to gain trust and steal sensitive information. The goal isn't just data theft; it's the infiltration of systems, the potential for espionage, or even the ability to manipulate critical infrastructure. Consider the impact on a military's communication networks, disrupted by a carefully crafted phishing operation, a testament to the power of human error.

Malware, a digital Trojan horse, silently infiltrates systems, often exploiting zero-day vulnerabilities—software flaws unknown to developers. This hidden threat can steal data, control systems, or even destroy them entirely. The aftermath of a successful malware campaign can be catastrophic, impacting not only computer systems but entire organizations, even governments, underlining the pervasive nature of cyber threats.

Data manipulation and disinformation campaigns, less about technical prowess and more about strategic deception, can be remarkably effective. Think of a carefully orchestrated campaign to sow discord, eroding public trust in institutions or inciting social unrest. This isn't about hacking systems; it's about manipulating perceptions and undermining stability, leveraging the power of information as a weapon.

Advanced persistent threats (APTs), subtle but dangerous intrusions into networks, often go unnoticed for extended periods. They operate silently, slowly collecting data or establishing footholds for future attacks. The discovery of an APT isn't merely about data loss; it's about the potential for long-term espionage, sabotage, and theft of intellectual property. Imagine the implications for a defense contractor whose designs have been stolen, a consequence of undetected persistent threats.

Exploiting vulnerabilities in industrial control systems (ICS) offers the potential for widespread physical damage. A cyberattack targeting a power plant, water treatment facility, or transportation network could cause devastating real-world consequences. Consider the scenario of a compromised dam, potentially unleashing a catastrophic flood; this emphasizes the physical impact of cyber warfare's destructive potential.

Supply chain attacks, focusing on the vulnerability of software and hardware supply chains, are a particularly insidious threat. By compromising a trusted vendor, an adversary can infect countless systems throughout a vast network. Imagine a malicious chip implanted in a server, potentially compromising countless sensitive systems. This demonstrates the cascading nature of supply chain attacks, a devastating form of strategic infiltration.

Network infiltration, the act of penetrating network security perimeters, allows access to valuable data and internal systems. The methods used range from sophisticated hacking techniques to exploiting human error. Once inside, the attacker can move laterally, gaining access to ever more sensitive information. Think of the damage caused by a breach into a bank's internal systems, emphasizing the financial and reputational impact of successful network infiltration.

Digital camouflage, in the cyber realm, aims to obscure an attacker's online presence. This involves masking IP addresses, using anonymizing tools, and employing other techniques to make attribution extremely difficult. The effectiveness of these techniques highlights the challenge of tracing cyberattacks and holding those responsible accountable.

Finally, understanding and mitigating the risks inherent in cyber warfare requires a multi-layered approach. This includes robust cybersecurity measures, employee training to identify and avoid phishing attacks, and the development of incident response plans for when attacks do occur. A proactive, comprehensive strategy is crucial for protecting valuable assets and maintaining operational continuity in the face of the ever-evolving cyber threat landscape. It emphasizes the importance of a holistic approach, blending technical security with human awareness and robust response plans.

4.2 Digital Camouflage

Digital camouflage isn't just about painting a tank green; it's a multifaceted strategy demanding a profound understanding of the digital battlefield. Think of it as the art of disappearing in plain sight, not through physical invisibility, but through the calculated manipulation of data and online presence. It's about controlling the narrative, not just concealing your actions. Mastering this requires a blend of technical skill and strategic thinking, a delicate balance between obfuscation and deception.

Effective digital camouflage starts with understanding your adversary's methods. What information are they seeking? What techniques do they typically employ for surveillance and reconnaissance? The answers to these questions inform your strategy. You need to know your enemy's methods to counter them effectively. It's a cat-and-mouse game, and you must be one step ahead. Consider, for instance, the use of honeypots—decoys designed to lure attackers away from critical systems. These are invaluable tools, drawing the enemy's attention while safeguarding your true digital assets. Imagine them as digital tripwires, signaling an intrusion attempt while providing valuable intelligence on your attacker's tactics.

Next, consider the importance of data encryption and anonymization. This isn't just about using strong passwords; it's about layering security protocols, using virtual private networks (VPNs) and proxies to mask your IP address, and encrypting all sensitive data both in transit and at rest. This is the foundation of your digital camouflage, the invisible armor protecting your digital assets from unwanted eyes. Imagine it like a multi-layered shield, constantly adapting and evolving to deflect incoming attacks. Don't just rely on one security measure; building a robust system requires multiple layers of defense.

Beyond technical measures, remember the human element. Social engineering is a potent weapon in any asymmetric conflict, and the digital world offers countless avenues for exploitation. False identities, fabricated online personas, and carefully crafted misinformation campaigns can all be valuable tools. Think of it as digital misdirection, creating a smokescreen that conceals your true objectives and activities. It's about manipulating the perception of your online presence to your advantage, often through carefully constructed stories and disinformation campaigns.

Furthermore, a key element of digital camouflage is the ability to adapt and evolve. The digital landscape is constantly shifting, and what works today might not work tomorrow. Regular security audits, software updates, and vigilance against emerging threats are crucial. Your digital camouflage must be dynamic, constantly adapting to counter new threats and maintain its effectiveness. Imagine it as a living organism, constantly evolving and adapting to survive.

Equally important is the need for compartmentalization. Divide your digital infrastructure into smaller, isolated units, so that a compromise in one area doesn't expose your entire system. This principle of compartmentalization is fundamental to both operational security and minimizing the damage of any successful attack. Consider it like a network of independent fortresses, each one capable of defending itself against intrusion, ensuring a breach in one area won't compromise the others. This compartmentalized approach is crucial in minimizing damage if your security is breached.

Finally, understand that digital camouflage is not a one-time solution but a continuous process. It requires constant vigilance, adaptation, and a proactive approach to security. Think of it as a marathon, not a sprint. It demands constant monitoring and a commitment to staying ahead of the curve. The threat landscape is always changing, and to stay ahead, you must constantly adapt and improve your digital camouflage. The aim is not just to hide, but to mislead and control the information your adversary receives. This ongoing vigilance is the key to maintaining your position of strength and effectiveness in the digital world.

Ultimately, mastering digital camouflage in the realm of asymmetric conflict demands a holistic approach that incorporates technical proficiency, strategic thinking, and a deep understanding of human behavior. It's not about hiding completely; it's about controlling the narrative, manipulating perceptions, and maintaining a decisive advantage. It's about creating ambiguity and deception in the digital landscape, giving you the freedom to operate while maintaining the element of surprise. Only then will your digital presence become an unreadable cipher, a true form of digital camouflage.

5 Chapter 5: Case Studies: Modern Conflicts

The Iraqi invasion of Kuwait in 1990, a blatant act of aggression, showcased a different kind of deception. Saddam Hussein's regime, despite its overwhelming conventional military might, employed a sophisticated strategy of misinformation, aiming to downplay its true intentions and sow confusion amongst its adversaries. Intelligence agencies struggled to fully decipher the Iraqi regime's plans, highlighting the efficacy of deception even in the face of significant intelligence gathering. The successful deception contributed significantly to the swift initial success of the Iraqi invasion. The subsequent international response, however, serves as a stark reminder that even the most elaborate plans can be undone by a united and decisive counteraction. The deception worked initially, yet ultimately failed in achieving its long-term strategic goals.

The protracted Afghanistan conflict, spanning decades, offers a compelling study in asymmetric warfare. The Taliban, significantly outmatched in conventional military capabilities, expertly utilized deception and concealment to sustain their insurgency. Their mastery of blending into the local population, coupled with effective propaganda campaigns aimed at influencing global perceptions, effectively prolonged the conflict for years. The physical landscape provided unparalleled cover, allowing for guerilla tactics and the creation of safe havens. Effective use of terrain and local knowledge allowed the insurgents to effectively counter Western advances. The conflict serves as a stark reminder of the limitations of superior military technology when faced with an adaptive and determined enemy skilled in the art of deception and concealment. The outcome underscores the importance of understanding cultural contexts and the limitations of purely military solutions in asymmetric conflicts.

The ongoing conflict in Ukraine provides a fascinating, contemporary case study. Russia's initial invasion, characterized by a combination of overt military action and covert disinformation campaigns, aimed to swiftly seize control. The deception involved multiple facets, from misrepresenting the scale of the invasion force to spreading false narratives about Ukrainian intentions. However, Ukraine's effective use of social media and open-source intelligence countered this misinformation campaign, limiting the impact of the initial deception attempt. The initial Russian successes were hampered by the unexpected Ukrainian resistance, fueled by both national spirit and access to timely information. The ongoing conflict highlights the critical role of information warfare and rapid information dissemination in both deception and its counteraction within the modern battle space, and reveals the limitations of outright military strength in the face of determined and adaptable opposition. The unexpected resilience of the Ukrainian forces, strengthened by a potent information war, provides a counterpoint to the narrative of rapid, decisive military victory through subterfuge. This ongoing conflict serves as a potent and evolving case study, demanding continuous analysis and re-evaluation of the effectiveness of deception in the modern era. The adaptability of both sides and the rapidly changing information landscape highlight the complexities and continuous evolution of conflict in the 21st century. The Ukrainian conflict underscores the crucial need for adaptability and the importance of understanding information warfare, not merely for its deception potential, but also for the ability to counteract such tactics effectively.

The three conflicts—the Gulf War, the Afghanistan conflict, and the ongoing situation in Ukraine—represent diverse but interconnected examples of deception and concealment in modern asymmetrical warfare. Each reveals unique strategies, showcasing the adaptive nature of warfare and the constantly evolving tactics employed by both belligerents and victims. These case studies, though seemingly disparate, collectively highlight the enduring importance of understanding and countering deception in any conflict environment, underlining the critical need for strategic awareness and information dominance.

5.1 Gulf War Deception

The 1990-1991 Gulf War presented a fascinating case study in deception, highlighting both the effectiveness and limitations of such strategies in a high-stakes conflict. The Iraqi regime, under Saddam Hussein, employed a variety of deceptive tactics, ranging from the strategic misdirection of troop deployments to the dissemination of false intelligence regarding its weapons capabilities. These efforts, though significant, ultimately fell short of their intended objective, a testament to the complexities of deception on a large-scale conflict.

A crucial element of Iraq's deception plan involved concealing the true extent and nature of its military buildup along the Kuwaiti border. Satellite imagery, while providing some valuable information, was deliberately obfuscated by Iraqi forces through the use of camouflage and the strategic positioning of equipment. This active concealment created ambiguity, making it difficult for coalition forces to accurately assess Iraq's military posture and intentions. The deception, therefore, played a significant role in delaying the initial coalition response, buying Iraq valuable time.

However, the Iraqi deception efforts suffered from key vulnerabilities. Human intelligence, obtained through defections and intercepted communications, exposed some of the inaccuracies in Iraq's public pronouncements. Moreover, the sheer scale of Iraq's military deployment, though camouflaged, eventually became increasingly difficult to entirely hide. This meant that, while the deception might have worked initially, delaying immediate and decisive action, it proved incapable of concealing the totality of their military preparations.

The information operations employed by Iraq also proved to be a double-edged sword. While some disinformation campaigns successfully sowed confusion among coalition forces, others backfired spectacularly. The blatant exaggerations and outright fabrications employed by Iraqi propaganda outlets undermined their credibility, particularly within international circles. This erosion of trust, inadvertently, heightened the coalition's resolve and fueled their determination to counter Iraq's aggression.

The use of psychological warfare techniques by Iraq was similarly flawed. Attempts to intimidate neighboring countries and instill fear amongst the populace proved relatively ineffective. Rather than fostering paralysis, these actions galvanized regional and international opposition, bolstering the coalition's legitimacy and support. The overall message, therefore, was that while deception can be a potent tool, its efficacy is highly contingent on its sophistication, credibility, and consistency.

Crucially, the Gulf War also highlighted the importance of accurate intelligence gathering and analysis in countering deceptive tactics. The coalition's success in penetrating Iraq's deception stemmed in large part from its superior intelligence capabilities. This highlighted the significant role of human intelligence, signal intelligence, and technological advancements, such as satellite imagery analysis, in unraveling deception strategies and understanding adversary's capabilities.

Ultimately, while Iraq's deception strategy partially succeeded in delaying the coalition response and obscuring some details of their military preparations, it ultimately failed to achieve its broader strategic objectives. The inherent limitations of the deception, coupled with the successes of the coalition's intelligence efforts, exposed its vulnerabilities. The case serves as a stark reminder of the crucial need for sophisticated intelligence-gathering and analytical capabilities in navigating the complex landscape of asymmetrical warfare and countering the use of deception. Successful deception requires not only skillful implementation but also a comprehensive understanding of the adversary's intelligence capabilities and methods. It also underlines the significance of maintaining the credibility of one's own communication and information channels, which can significantly impact the credibility of one's deceptive actions and strategies.

5.2 Afghanistan Conflict Analysis

The Soviet-Afghan War, and its aftermath, offer a compelling case study in deception and concealment within an asymmetric conflict. The mujahideen, vastly outgunned by the Soviet military, relied heavily on unconventional tactics, often exploiting the terrain and local knowledge to their advantage. Their success wasn't simply a matter of superior guerilla warfare, however; a significant aspect involved meticulous deception, creating a fog of war that hampered Soviet efforts. Ambushes were carefully planned and concealed, drawing the Soviets into deadly traps. The seemingly random attacks were, in fact, meticulously orchestrated, leveraging information gleaned from informants and exploiting the vulnerabilities of the Soviet supply lines.

Communication, or rather the lack thereof, played a crucial role. The mujahideen skillfully used the mountainous landscape to their advantage, communicating through established, yet clandestine, networks. This intricate network of informants and couriers allowed for the rapid dissemination of information and coordinated attacks, while simultaneously concealing their true strength and intentions from Soviet intelligence. They mastered the art of disappearing into the landscape, making it almost impossible for the Soviets to locate and engage them effectively. This was achieved not only through physical concealment but also through psychological manipulation, creating a sense of constant uncertainty and fear amongst the Soviet troops.

The use of propaganda, another key element in this asymmetrical war, also reveals an adept use of deception. Information circulated – both internally within the mujahideen ranks and externally to international audiences – strategically inflated their numbers and military capabilities. This wasn't about outright lies but the manipulation of truth, amplifying successes while minimizing losses, a method of deception that proved profoundly effective in garnering international support and prolonging the conflict. This amplified the perception of a far greater threat than what the Soviets were actually facing. The Soviet Union found themselves entangled in a conflict that was difficult to define, let alone win.

The post-Soviet period saw a continuation of these deception tactics, albeit in a different form. The Taliban's rise to power was in no small part due to their ability to exploit the power vacuum and manipulate the local population. They adeptly played on grievances and utilized propaganda to gain legitimacy, cultivating support among tribal leaders and influencing the general populace. This was a form of deception subtly woven into the fabric of Afghan society, where trust and loyalty shifted according to strategic needs. Their early strategy focused on gradual infiltration, gradually extending their control over various regions through a combination of force, negotiation, and strategic alliances. Their resilience and ability to adapt – often blending into the civilian population – made them an elusive enemy.

Understanding the specific nature of deception and concealment in Afghanistan requires an examination of the cultural context. Afghan society, historically marked by tribal allegiances and a complex power dynamic, presents fertile ground for deception and strategic manipulation. Relationships are fluid, loyalties often shifting based on perceived advantage. These fluid social structures allowed the various factions, including the mujahideen and the Taliban, to blend effectively within the population, rendering the distinction between civilian and combatant at times blurred, especially for external observers. This was a critical advantage, allowing combatants to move freely, gather intelligence, and launch attacks with relative impunity. Foreign forces often struggled to distinguish friend from foe, further highlighting the efficacy of this cultural camouflage.

The use of improvised explosive devices (IEDs), a hallmark of the insurgency in Afghanistan, provides further evidence of successful deception. Their seemingly random deployment created a pervasive sense of danger, disrupting the flow of operations and crippling logistical efforts. Often, these IEDs were effectively concealed, making detection extremely challenging for even the most sophisticated countermeasures. These concealed weapons were often planted by seemingly ordinary individuals, blurring the lines between civilian and militant further increasing the effectiveness of the overall deception strategy.

Therefore, the Afghan conflict stands out as a stark example of how deception and concealment, intertwined with cultural context and leveraged through effective exploitation of terrain and intelligence, can shape the outcome of an asymmetric conflict. The success of insurgent groups in Afghanistan serves as a testament to the importance of understanding the human factor and the psychological aspects of warfare, factors often overlooked in traditional military planning. The lessons from Afghanistan underscore the need for adaptable strategies that account for the nuanced realities of asymmetric conflicts. The enduring legacy of the conflict is not merely a record of military engagements but a masterclass in strategic deception and its enduring potency.

5.3 Ukraine Conflict Insights

The Ukrainian conflict presents a fascinating, albeit tragic, case study in modern asymmetrical warfare. Russia, possessing a significantly larger conventional military, initially employed a strategy of swift, decisive action, aiming for a rapid overthrow of the Ukrainian government. This was predicated on a gross miscalculation, a profound underestimation of Ukrainian resolve and the effectiveness of Western support. The initial blitzkrieg faltered, encountering unexpectedly fierce resistance. This failure exposed the limitations of sheer military might when facing a determined opponent utilizing effective asymmetric tactics.

This initial failure forced a recalibration of Russian strategy.

The focus shifted from a rapid, decisive victory to a protracted war of attrition, leveraging superior firepower and manpower to gradually erode Ukrainian defenses. This transition, however, highlighted another crucial element of asymmetric warfare: the importance of information operations. Russia's narratives about the conflict, often disseminated through state-controlled media and online propaganda, were designed to shape global perceptions, weaken international support for Ukraine, and sow discord within Ukrainian society. This was countered by Ukrainian efforts to maintain a robust flow of verified information and to expose Russian disinformation.

Concealment, in the context of the Ukrainian conflict, has played a vital, multi-faceted role. Ukrainian forces have effectively employed camouflage and concealment to ambush Russian troops, hindering the advance of larger, less adaptable forces. Their mastery of operational security has proven crucial in sustaining effective resistance, while protecting key infrastructure from Russian attacks. The conflict also showcases the importance of network security and the challenges of securing critical infrastructure and communications against cyber attacks, a major facet of modern warfare. Russia has attempted to disrupt Ukrainian communications and critical infrastructure by employing a variety of cyber warfare tactics. Simultaneously, Ukraine has utilized cyber deception strategies to confuse and mislead the Russians. The use of drones, for instance, has proven exceptionally effective; relatively inexpensive yet highly impactful, providing a cost-effective counterbalance to superior Russian firepower. This exemplifies how smaller, less resourced forces can utilize effective asymmetric tactics to neutralize the advantages of a larger foe.

The conflict also highlights the crucial role of international support in asymmetrical warfare. The provision of weapons, training, and intelligence by Western nations has significantly enhanced Ukraine's capacity to resist. This external support demonstrates how the dynamics of an asymmetrical conflict can extend far beyond the immediate battlefield, influencing global power dynamics. It also underscores the vulnerabilities of a seemingly superior military force when confronted by a technologically advanced and politically supported opponent utilizing an effective strategy based on asymmetric principles.

Furthermore, the Ukrainian conflict demonstrates the fluidity of modern warfare. The initial Russian strategy, based on a conventional approach, failed to account for the resilience of the Ukrainian people and the complexities of modern information warfare. The ensuing shift to a war of attrition has seen a corresponding increase in reliance on information and cyber operations, highlighting the intertwined nature of these facets of modern conflict. The conflict has demonstrated the importance of adapting and evolving tactics in response to changing circumstances. The ongoing conflict continues to highlight the ongoing necessity of refining and adapting strategies to the challenges of modern warfare.

Finally, the conflict serves as a powerful reminder of the limitations of military power alone. While Russia's superior military capabilities have been undeniable, they have been unable to guarantee a swift victory. The Ukrainian resistance, fueled by national unity and international support, has demonstrated the enduring power of asymmetric warfare in challenging conventional military might, shaping the conflict's trajectory in ways that were unforeseen at its outset. The conflict serves as a testament to the vital role of strategy, adaptation, and the relentless pursuit of unconventional advantage in asymmetrical conflict. It also showcases the critical importance of understanding the human element within any conflict – the unwavering resolve of the defenders and the capacity of even a numerically inferior force to inflict significant damage on its larger opponent. The resilience and creativity displayed by Ukraine underscores the unexpected twists and turns that can define asymmetrical conflicts, making them difficult to predict and control.

6 Chapter 6: Future of Asymmetric Warfare

The proliferation of readily available, sophisticated technologies will fundamentally reshape asymmetric warfare. No longer the sole domain of state actors, advanced drones, readily accessible encryption software, and even AI-powered analytical tools are empowering non-state actors to challenge conventional forces in unprecedented ways. This democratization of technological power necessitates a re-evaluation of traditional defense strategies and a focus on adaptability and agility rather than sheer firepower. We're entering an era where the underdog can level the playing field through clever application of technology.

Consider the potential impact of readily available commercial satellite imagery. Previously, access to such data was restricted to powerful governments, providing them with a significant intelligence advantage. Now, anyone can purchase high-resolution imagery, enabling non-state actors to monitor military movements, identify vulnerabilities, and even plan attacks with remarkable precision. This fundamentally alters the equation, blurring the lines between state and non-state actors and challenging traditional power dynamics.

The future of asymmetric warfare will likely be characterized by a blend of highly technological and low-tech approaches.

We might see a resurgence in unconventional tactics, such as guerilla warfare, but augmented by the precision and efficiency afforded by technology. Think of coordinated attacks enabled by encrypted communication networks, using commercial drones for reconnaissance and even small-scale attacks, all planned and executed with precision guided by readily accessible intelligence. The lines are increasingly blurred between what is considered a "high-tech" and "low-tech" approach in this context.

This blurring of lines requires a more nuanced approach to defense. Traditional military superiority might be less effective against networked, decentralized adversaries who leverage technology strategically. We must shift away from a focus solely on superior weaponry and towards a multi-layered defense system that emphasizes information security, resilient infrastructure, and rapid adaptation to changing threats. The speed at which technology evolves demands agility in our defensive strategies.

Moreover, the ethical implications are profound. The easy accessibility of advanced technologies could lead to an escalation of conflict and an increase in civilian casualties. The line between legitimate self-defense and indiscriminate violence becomes increasingly blurred as the playing field levels, requiring an urgent re-evaluation of international law and norms. The question of attribution and accountability in the digital realm also presents a significant challenge. We need to develop international mechanisms for accountability, in order to mitigate the risks associated with the increasing use of technology in asymmetric conflicts.

The future of asymmetric warfare is not simply about more powerful weapons; it's about adaptability, intelligence, and a profound understanding of the adversary's methods and motivations. It's about understanding the intricate interplay between technology, human factors, and the geopolitical context in which the conflict unfolds. We must strive to anticipate and counter the changing tactics, not only by developing countermeasures but also by understanding the psychological and sociological factors which drive asymmetric conflict in the first place. The future demands a more holistic approach, one that encompasses not only technological advancements but also socio-political and psychological considerations. This will be pivotal in navigating the evolving landscape of future conflicts. By anticipating the future developments and adapting our strategies accordingly, we can better prepare for and navigate the complexities of asymmetric warfare in the years ahead. Predicting the future is inherently uncertain, but by recognizing the trends outlined above, we can develop more robust and effective strategies for mitigating the threats and challenges posed by asymmetric warfare. It demands a collaborative approach, involving policymakers, military strategists, technologists, and social scientists working together to foster proactive measures, mitigate the risks of escalation, and establish a framework for ensuring the responsible use of technology in conflicts, all while keeping human rights and international law at the forefront of our strategic thinking. The stakes are high, and the need for creative solutions is urgent.

6.1 Technological Advancements

The proliferation of readily available drone technology significantly alters the asymmetric warfare landscape. No longer the exclusive domain of state actors, inexpensive, commercially available drones offer a potent tool for non-state actors, enabling reconnaissance, surveillance, and even limited attacks with devastating precision. This accessibility democratizes warfare, leveling the playing field, at least in terms of reconnaissance capabilities, but also introduces new challenges for conventional forces. The sheer volume of potential drone deployments makes traditional countermeasures, like jamming, difficult to implement effectively and economically. We are entering an era where even a small insurgent group can utilize a swarm of drones to achieve disproportionate effects, overwhelming defenses through sheer numbers and distributed attacks. The potential for creating chaos and disruption is undeniable. Consider, for example, a scenario where a coordinated drone swarm disables critical infrastructure, crippling power grids or communication networks, thereby gaining a substantial strategic advantage in the asymmetry of warfare. This presents a difficult challenge for the established, conventional powers and forces, who may struggle to maintain superiority in this ever changing environment.

Developing effective counter-drone technologies is paramount. This isn't merely a matter of developing more advanced jamming systems or deploying anti-drone weaponry; it's about developing a layered, adaptable defense capable of responding to rapidly evolving threats. Imagine the difficulties faced by a military struggling to protect its bases and supply lines from constant drone attacks, each wave more sophisticated and unpredictable than the last. This demand places an even greater emphasis on intelligence gathering and predictive modeling. Understanding the capabilities and intentions of potential adversaries, including their likely technological acquisitions and deployment strategies, is crucial to developing effective countermeasures. Effective response hinges not only on technological advancement but also on proactive intelligence gathering and swift adaptation. It's a race to stay ahead of an increasingly inventive opponent.

Beyond drones, advancements in artificial intelligence (AI) pose both threats and opportunities. AI-powered algorithms are rapidly improving at tasks crucial to both deception and concealment, from sophisticated data analysis for identifying patterns of enemy movement to real-time adaptation of camouflage patterns based on environmental changes. AI can even improve the capabilities of deception, producing more convincing disinformation campaigns, tailoring messages for maximum impact and manipulating sentiment analysis. However, these same AI advancements can assist in defensive measures. For example, AI could improve the accuracy and speed of threat detection and identification, allowing for a quicker and more decisive response. The ability of AI to analyze vast datasets and identify subtle anomalies could provide invaluable insights that would be impossible for humans to discern without the help of AI. The key lies in understanding the dual-use nature of this technology and harnessing its capabilities for both offense and defense.

The implications for operational security are profound. In an increasingly interconnected world, the risk of information leaks, regardless of how insignificant they may appear, is ever present. Sophisticated cyberattacks can exploit vulnerabilities, compromising sensitive data and potentially undermining entire military operations. This requires a shift in mindset, emphasizing proactive risk mitigation and the implementation of robust cybersecurity measures. It's more than just updating software; it's about establishing a security culture that permeates every level of an organization. This security culture needs to include comprehensive training programs, rigorous testing, and continuous monitoring, always anticipating and adapting to emerging cyber threats, especially considering the increased rate of technological innovation in a war environment. Effective operational security in the age of widespread technological proliferation needs to be both highly technologically sophisticated and completely

integrated into the organization's culture to be effective.
Finally, the ethical considerations surrounding the use of advanced technologies in asymmetric warfare are paramount. The potential for unintended consequences, civilian casualties, and the escalation of conflict is significant. The development and deployment of autonomous weapons systems, for example, raise profound ethical questions about accountability and the potential for machines to make life-or-death decisions without human oversight. A thorough and ongoing ethical discussion is necessary, guided by international law and humanitarian principles, to ensure that technological advancements are used responsibly and ethically, minimizing the potential for harm and fostering accountability in the battlefield. The future of asymmetric warfare will hinge not only on technological superiority but also on ethical considerations and the willingness to engage in responsible innovation.

6.2 Evolving Tactics

The proliferation of readily available drone technology is fundamentally altering the landscape of asymmetric warfare. No longer the exclusive domain of nation-states, these relatively inexpensive unmanned aerial vehicles (UAVs) empower non-state actors with previously unimaginable capabilities, dramatically shifting the balance of power. This accessibility democratizes surveillance, reconnaissance, and even limited-scale attacks, forcing a reevaluation of traditional defense strategies. The ease of acquiring and deploying these systems presents a significant challenge, demanding innovative countermeasures.
Simple, readily available counter-drone technologies aren't always enough. Sophisticated jamming or directed energy weapons might be too expensive or require specialized training

for effective deployment in many scenarios. Therefore, focusing on deception and concealment strategies becomes paramount. The key lies in creating a multi-layered defense, combining technological solutions with human ingenuity and adaptable tactics. One effective approach involves developing decoy systems—simulating critical infrastructure or troop movements using easily produced, cost-effective, and easily deployed imitations. This requires a deep understanding of an adversary's technological capabilities and their patterns of information gathering. By strategically deploying these decoys, one can divert resources and attention away from genuine targets, creating opportunities for counter-offensive actions or buying valuable time for a more robust response. The effectiveness of this strategy rests on understanding the adversary's operational patterns and utilizing this knowledge to build credibility within the deception.

Furthermore, exploiting the limitations of readily-available drone technology is vital. Many commercially available drones rely heavily on GPS for navigation. Therefore, employing countermeasures that disrupt or spoof GPS signals can render these systems ineffective or even cause them to self-destruct. Furthermore, the limited range and payload capacity of many consumer-grade drones offer an inherent vulnerability. Understanding these limitations allows for the creation of robust countermeasures, minimizing the threat these systems pose.

Beyond technological countermeasures, the human element remains crucial. Training personnel to recognize and effectively counteract drone threats, including recognizing patterns of surveillance and developing robust operational security (OPSEC) procedures, is non-negotiable. This includes educating personnel on identifying potential drone activity, responding to sightings, and reporting suspicious activity to appropriate channels.

The effectiveness of any evolving tactic relies on adaptability. As drone technology continues to evolve, so must our methods for countering it. A flexible approach, one that embraces continuous assessment, refinement, and adaptation, is critical in maintaining a decisive advantage. Regularly updating counter-drone measures, incorporating lessons learned from past encounters, and investing in research and development of new strategies are essential for adapting to this ever-shifting landscape. Only by embracing constant evolution can we successfully navigate the complexities of this new battleground.

The challenge of combating the proliferation of drone technology in asymmetric warfare demands a holistic approach. It's not merely about acquiring advanced technology; it's about strategic thinking, resourcefulness, and adaptable tactics. Combining technological countermeasures with comprehensive training programs, robust OPSEC protocols, and an emphasis on creative deception will be crucial in maintaining a decisive edge against this pervasive threat. The future of effective counter-drone strategies lies in developing layered defenses that combine cutting-edge technology with innovative tactical thinking. The focus needs to shift from solely reactive measures to proactive, adaptive strategies. This requires a continuous cycle of assessment, adaptation, and innovation to maintain effectiveness in the face of an evolving threat.

7 Chapter 7: Conclusion: Adapting to Deception

The persistent evolution of asymmetric warfare necessitates a fundamental shift in our approach to defense. We can no longer rely on solely technological superiority; instead, we must cultivate a deeper understanding of deception's multifaceted nature and its implications for strategic decision-making. This understanding must permeate every level of military planning, from individual soldier training to national-level strategy. Failure to adapt will leave us vulnerable to increasingly sophisticated forms of deception, leading to potentially catastrophic consequences.

Adapting to this evolving landscape requires a proactive, multilayered strategy. Investing in advanced detection and analysis technologies is crucial. These tools, however, are only as effective as the human analysts interpreting their output. Therefore, rigorous training programs focusing on critical thinking, pattern recognition, and the identification of subtle deception cues are paramount. This includes understanding the psychological motivations behind deceptive tactics, allowing for more effective countermeasures. Our focus must be on cultivating skepticism and fostering a culture of questioning assumptions.

Beyond technological advancements and human expertise, ethical considerations demand careful attention. The use of deception, while often necessary, carries inherent risks. The potential for unintended consequences, escalation, and the erosion of trust must be carefully weighed against the strategic benefits. Clear guidelines and robust oversight mechanisms are essential to ensure the responsible application of deception techniques. Transparent internal review processes, combined with open dialogue regarding the ethical implications of specific tactics, are crucial components of maintaining a responsible defense posture.

The strategic implications of widespread deception are far-reaching. It requires a paradigm shift in intelligence gathering, analysis, and dissemination. Traditional methods are often insufficient to uncover complex deception campaigns. We need to develop innovative strategies that focus on verifying information from multiple, independent sources. This includes strengthening relationships with allied intelligence agencies and fostering collaborative efforts to share information and collectively develop counter-deception strategies. Such collaborations must be based on mutual trust and transparency, which may require revisiting some outdated intelligence sharing protocols.

The future of defense rests on our ability to anticipate, adapt, and effectively counter deception. This requires a multifaceted approach that combines technological innovation with human ingenuity, ethical considerations with strategic acumen. The challenge is not merely to detect deception, but to anticipate its forms and disrupt its effectiveness. The ability to anticipate an adversary's deceptive strategies allows for proactive measures, reducing the likelihood of surprise and limiting the impact of potentially devastating attacks. Our success lies not in technological superiority alone, but in the development of a resilient and adaptive security posture that anticipates and effectively counters the ever-evolving challenges of asymmetric conflict. This demands continuous learning, open communication, and a willingness to adapt and evolve alongside the landscape of deception itself. Ignoring this reality presents significant risks to national security, and only the most agile and forward-thinking organizations will ultimately thrive in this new age of warfare. The fight against deception is a constant, ongoing process that demands a commitment to lifelong learning and an embrace of change. The future of defense, therefore, lies in our collective ability to remain vigilant and adaptable, constantly refining our understanding and our strategies to meet the ever-changing threat. The fight against deception is not just about winning battles but about safeguarding the future.

7.1 The Future of Defense

The increasing reliance on technology necessitates a fundamental shift in defensive strategies. No longer can we rely solely on traditional methods; the battlefield of tomorrow will be defined by the seamless integration of physical and digital domains. This demands a multi-layered approach, combining robust physical security measures with equally sophisticated cyber defenses. Think of it as a fortress, not just with strong walls, but also with an impenetrable digital shield.

Developing this integrated defense system requires a comprehensive understanding of the enemy's capabilities. This includes not just their conventional weaponry, but their potential for cyberattacks, disinformation campaigns, and the exploitation of vulnerabilities in critical infrastructure. Intelligence gathering and analysis become paramount, allowing for proactive threat assessment and the preemptive neutralization of hostile actions. We must move beyond reactive measures and embrace proactive defense, anticipating and mitigating threats before they materialize.

Furthermore, the future of defense hinges on cultivating adaptability and resilience. The speed of technological advancement means that any static strategy will quickly become obsolete. We must foster a culture of continuous learning and innovation within our defense forces, embracing new technologies and adapting our tactics to counter emerging threats. This requires significant investment in research and development, focused on creating adaptable systems capable of evolving alongside the threats they confront. Investing in education and training is paramount as well, to ensure our personnel are equipped to operate and maintain these new technologies effectively.

Success in future conflicts will likely depend on the ability to leverage artificial intelligence effectively. AI can aid in analyzing vast amounts of data to identify patterns and predict enemy actions, giving us a crucial advantage in intelligence gathering and decision-making. AI-powered systems can also assist in the development of more sophisticated camouflage and concealment techniques, making our forces harder to detect and track. But remember, the reliance on AI also creates vulnerabilities; robust cybersecurity protocols are crucial to prevent adversarial manipulation of these systems. Our reliance on AI should always be balanced by the retention of human oversight and critical thinking.

Effective deception will play a vital role in future conflicts. However, this doesn't mean resorting to outright misinformation. Instead, it involves creating ambiguity and uncertainty in the minds of our adversaries. By carefully managing the information we release and controlling the narrative, we can manipulate their perceptions of our capabilities and intentions, making it difficult for them to anticipate our moves. The aim here is not to lie, but to effectively utilize ambiguity. We need to carefully consider the ethical implications of deception in war, and it must never cause harm to non-combatants.

Finally, international cooperation and information sharing are indispensable. No single nation can effectively address the complex challenges of future warfare alone. Collaborating with allies allows for the pooling of resources, the sharing of intelligence, and the development of joint strategies to counter emerging threats. This interconnected approach, fostering mutual trust and coordinated responses, forms a vital component of the long-term success and peace in the future. Strengthening alliances and creating networks for information exchange are crucial for mitigating these future challenges. The future of defense isn't about any one nation, it's about cooperation to secure lasting peace.

7.2 Ethical Considerations

The inherent tension between military necessity and ethical conduct is sharply defined in asymmetric warfare. While deception and concealment are often vital for survival and success, their employment demands careful consideration of the moral implications. The line between justifiable strategy and unacceptable brutality can be perilously thin, particularly when dealing with non-state actors or in conflicts where civilian populations are heavily involved. This necessitates a rigorous ethical framework guiding the planning and execution of any deceptive strategy.

A key ethical concern involves the potential for collateral damage. Deceptive tactics, such as disinformation campaigns, can unintentionally escalate conflict or spread misinformation that harms innocent individuals. The deliberate targeting of civilians, even indirectly through deceptive maneuvers, is unequivocally unethical and should be strictly avoided. Minimizing harm to non-combatants must remain a paramount concern. Careful risk assessments, meticulous planning, and constant monitoring of the situation are vital to mitigate such risks.

Furthermore, the use of deception raises questions of transparency and accountability. While some level of secrecy is inherent in military operations, a lack of transparency can erode public trust and fuel resentment. Governments and militaries must balance the need for secrecy with the need for accountability, striving to maintain public confidence even while employing deceptive strategies. This requires open communication where possible, and a robust system of internal oversight to ensure that deceptive tactics remain within ethical boundaries.

Proportionality is another critical ethical consideration. The level of deception employed should be proportionate to the threat posed. Employing elaborate and potentially harmful deceptive maneuvers against a relatively minor threat is unethical. A thorough assessment of the threat, factoring in the potential consequences of both employing and withholding deception, should always precede action. The potential for escalation should be weighed against the possible benefits of using deception.

The ethical dilemmas are amplified when dealing with non-state actors. Defining the boundaries of engagement and determining the appropriate level of deception become far more complex when facing adversaries who operate outside the established rules of conventional warfare. International humanitarian law should serve as a guideline; however, its application can be challenging in asymmetric conflicts that blur the lines between combatants and civilians. Nevertheless, adherence to its principles remains crucial to maintaining a moral high ground and preventing unintended consequences. Robust mechanisms for investigating alleged war crimes are crucial for upholding international legal standards.

Moreover, the rapid advancements in technology, particularly in areas such as cyber warfare and information operations, introduce new ethical challenges. The anonymity and scale afforded by digital technologies can facilitate the spread of disinformation and cyberattacks with potentially devastating effects. Developing effective and ethically sound countermeasures is critical to mitigate the risks associated with the use of deception in the digital realm. Strict adherence to international norms and agreements governing cyber warfare is vital.

In conclusion, the ethical considerations surrounding deception and concealment in asymmetric warfare are multifaceted and require constant attention. A robust ethical framework that prioritizes proportionality, accountability, transparency, and the minimization of harm to non-combatants is essential to navigate the complex moral landscape of this type of conflict. This requires not only strict adherence to international humanitarian law but also a commitment to ongoing reflection and adaptation to the ever-evolving technological and strategic landscape of modern warfare. Only through constant vigilance and a firm commitment to ethical principles can we ensure that the use of deception remains a tool of justifiable strategy rather than an instrument of cruelty.

7.3 Strategic Implications

Understanding the strategic implications of deception and concealment in asymmetric warfare requires a nuanced perspective, moving beyond simple tactical analyses. It demands a comprehensive understanding of the long-term consequences of employing, or being subjected to, these methods. The very nature of asymmetric conflict, with its inherent power imbalances, necessitates a deep consideration of these ramifications. The impact extends far beyond the immediate battlefield, influencing geopolitical dynamics, international relations, and the overall stability of regions.

Successful deception, while offering tactical advantages in the short term, can have unpredictable long-term consequences. Erosion of trust, for instance, is a significant byproduct. A state consistently employing deceptive tactics risks alienating its allies and undermining international cooperation, potentially isolating itself in times of genuine crisis. The loss of credibility can prove far more damaging than any tactical victory achieved through deceit. This loss of trust extends beyond governmental actors, too; affecting the perception of the state amongst its citizens.

Conversely, falling victim to sophisticated deception operations can have devastating repercussions. Misinformation campaigns, designed to manipulate public opinion and sow discord, can destabilize governments and societies, even without direct military intervention. The spread of false narratives can fuel internal conflicts, polarize populations, and severely hamper effective governance. Understanding the information ecosystem and identifying vulnerabilities is therefore crucial to mitigating these risks. Effective counter-intelligence measures are vital to avoid being manipulated.

Economic consequences also warrant careful consideration. The resources dedicated to developing and implementing deception strategies represent a significant opportunity cost. These investments could potentially have been diverted towards enhancing economic development or social programs. Similarly, the economic fallout from a successful deceptive attack, be it a cyberattack crippling infrastructure or a disinformation campaign damaging international trade, can be staggering. Economic stability is intrinsically linked to national security, and a failure to recognize this connection can be catastrophic.

The ethical dimension is paramount. While deception is an accepted aspect of warfare, the line between acceptable tactical maneuvering and morally reprehensible actions remains blurry. The potential for civilian casualties, the use of disinformation to incite violence, and the psychological manipulation of populations are all serious ethical concerns. A clear ethical framework governing the use of deception in conflict is essential to prevent abuses and uphold humanitarian principles. This framework, however, must also acknowledge that asymmetric warfare demands adaptability and creativity, even when it means confronting ethically complex situations.

Finally, the strategic implications extend to the future. The constant evolution of technology, particularly in the cyber domain, necessitates a proactive approach to defense. Anticipating future deceptive techniques and developing robust countermeasures are no longer luxuries but necessities. Investment in research and development, coupled with a commitment to ongoing training and education, are vital to maintaining a strategic edge in an increasingly complex and uncertain global landscape. Failure to adapt proactively can leave nations vulnerable to unforeseen threats and unexpected strategic setbacks. Therefore, ongoing assessment and adaptation are crucial to managing the strategic risks inherent in the asymmetric warfare landscape.

www.ingramcontent.com/pod-product-compliance
Lightning Source LLC
LaVergne TN
LVHW022023200926
848481LV00007B/2510
* 9 7 9 8 3 4 9 4 7 1 6 6 7 *